# GENETIC ALCHEMY

# Genetic Alchemy

MIGUEL SALAZAR

Copyright © 2025 by Miguel Salazar
All rights reserved. No part of this book may be reproduced in any manner whatsoever without written permission except in the case of brief quotations embodied in critical articles and reviews.
First Printing, 2025

# CONTENTS

~1~

Introduction to Genetic Alchemy

1

~2~

Human - Alien Hybridization

9

~3~

Genetic Engineering and Manipulation

17

~4~

Psychological Effects of Hybridization

25

~ 5 ~

# Ethical Implications of Alien Genetics
33

~ 6 ~

# Cultural Interpretations of Hybrid Beings
41

~ 7 ~

# Hybridization in Sci-Fi Literature and Film
49

~ 8 ~

# Historical Accounts of Alien Encounters
57

~ 9 ~

# Hybridization and Evolutionary Biology
65

~ 10 ~

# Spiritual Perspectives on Hybrid Existence
73

~ VII

~11~

# Government Conspiracy Theories Regarding Hybrids
81

~12~

# Medical Research on Hybrid Physiology
89

~13~

# Conclusion and Future Directions
97

# ~ 1 ~

# INTRODUCTION TO GENETIC ALCHEMY

### DEFINITION AND SCOPE OF GENETIC ALCHEMY

Genetic alchemy represents a complex intersection of genetics, extraterrestrial influence, and the human experience, primarily characterized by the manipulation of genetic material to create hybrid beings. This concept extends beyond traditional genetic engineering, delving into the implications of integrating alien DNA with human genetic structures. Such practices suggest not only an advancement in scientific capabilities but also the potential for radical changes in human evolution and identity. The term evokes both ancient alchemical traditions, which sought to transform base substances into gold, and modern scientific endeavors aimed at enhancing human capabilities through genetic modifications.

The scope of genetic alchemy encompasses a wide array of phenomena related to human-alien hybridization. It involves the examination of historical accounts of alien encounters where hybrid beings are purported to exist. These narratives often describe individuals who exhibit physical and psychological traits distinct from those of the general human population, suggesting a blend of extraterrestrial and human characteristics. UFO enthusiasts and conspiracy theorists alike have long posited that these hybrids serve as a bridge between two species, raising questions about their roles, purposes, and the potential for future interactions between humans and their alien counterparts.

In exploring the psychological effects of hybridization, the discourse shifts towards understanding the mental and emotional impacts on both hybrids and the humans involved in these encounters. Reports often indicate that hybrids experience unique psychological challenges, such as identity crises and alienation from both human and extraterrestrial communities. These experiences can lead to broader discussions about the nature of consciousness and the implications of sharing one's existence with a non-human lineage. Moreover, the psychological ramifications extend to those who believe they have been in contact with hybrids, influencing their perceptions of reality and their place within the universe.

The ethical implications of alien genetics cannot be overlooked, as they raise significant moral questions surrounding consent, identity, and the potential for exploitation. As humanity grapples with the moral landscape of genetic engineering, the integration of alien DNA complicates these debates further. Are hybrids entitled to the same rights as humans? What responsibilities do researchers and governments have towards these beings? Such inquiries compel society to critically evaluate the boundaries of scientific exploration and the ethical considerations that accompany the pursuit of knowledge in this uncharted territory.

Culturally, genetic alchemy resonates through various channels, including science fiction literature and film, where hybridization serves as a potent narrative device. These works often reflect societal fears and aspirations regarding the future of humanity and the potential for coexistence with alien intelligences. Historical accounts and government conspiracy theories add another layer to the discourse, suggesting that the existence of hybrids is not merely speculative but rooted in a reality that has yet to be fully acknowledged. This amalgamation of perspectives fosters a rich cultural dialogue, enabling a deeper understanding of what it means to be human in a universe that may harbor other sentient beings.

## HISTORICAL CONTEXT OF ALIEN ENCOUNTERS

The historical context of alien encounters has been shaped by a confluence of myth, folklore, and documented events that span centuries. Ancient civilizations often recorded their experiences with celestial beings, interpreting these encounters through the lens of their cultural beliefs and societal norms. From the chariots of fire described in biblical texts to the numerous accounts of gods descending from the heavens, early interactions with extraterrestrial entities were frequently framed within the context of divine intervention or spiritual guidance. These narratives laid the groundwork for modern interpretations of alien encounters, suggesting that humanity has long grappled with the implications of beings from beyond our world.

As the scientific revolution progressed, the understanding of the cosmos shifted dramatically. The advent of telescopes and advancements in astronomy invited a new wave of speculation regarding life beyond Earth. By the late 19th and early 20th centuries, literature began to reflect this intrigue, spurring the genre of science fiction that often explored themes of alien visitation and hybridization. Works such as H.G. Wells' "The War of the Worlds" not only entertained but also provoked critical thought about human identity and the potential for genetic manipulation. In doing so, these narratives established a cultural framework within which the concept of alien hybridization could be examined, merging fiction with emerging scientific theories of evolution and genetics.

The mid-20th century marked a pivotal moment in the perception of alien encounters, particularly with the rise of the UFO phenomenon. Reports of sightings and alleged abductions became increasingly prominent, fueled by both media coverage and the burgeoning field of ufology. During this period, claims of human-alien hybrids began to surface, often accompanied by detailed accounts of alleged medical examinations and genetic experimentation. These stories resonated with a growing public fascination with the unknown and the possibility of advanced extraterrestrial civilizations. The psychological effects of such encounters, including trauma and altered perceptions of reality, became a focal point for researchers and enthusiasts alike, prompting deeper inquiries into the nature of these experiences.

The ethical implications of alien genetic engineering have also emerged as a significant area of discussion within this historical context. As the boundaries between human and alien blur, questions regarding the morality of hybridization practices arise. Are these hybrids a new form of life deserving of rights, or are they merely products of a sinister agenda? These concerns are often exacerbated by government conspiracy theories that suggest a cover-up of alien interactions and secret experiments on human subjects. Such theories, while often speculative, reflect the anxieties surrounding the unknown and the potential ramifications of tampering with the fabric of human existence.

Culturally, the interpretation of hybrid beings has evolved, influenced by societal attitudes towards diversity, identity, and belonging. As humanity grapples with its own evolutionary trajectory, the concept of hybrids serves as a mirror reflecting our fears and aspirations. From the monstrous representations in early cinema to the complex, relatable characters found in contemporary science fiction, hybrids challenge the definitions of humanity and provoke discussions about what it means to be human. This evolution of thought underscores the importance of examining historical accounts of alien encounters, as they not only inform our understanding of potential future interactions but also shape the narrative surrounding our place in the universe.

# ~ 2 ~

# HUMAN - ALIEN HYBRIDIZATION

THE SCIENCE BEHIND HYBRIDIZATION

The concept of hybridization, particularly in the context of human-alien interactions, draws upon a complex interplay of genetic science, evolutionary theory, and sociocultural narratives. At its core, hybridization involves the mixing of genetic materials from two distinct species, resulting in offspring that may exhibit traits from both progenitors. This phenomenon has been documented in various scientific fields, including agriculture and animal breeding, but when applied to extraterrestrial entities, it raises a host of questions about the very nature of human identity and the parameters of life itself. The science behind hybridization suggests not only biological possibilities but also an intricate web of ethical dilemmas and psychological ramifications.

Genetic engineering techniques, such as CRISPR and gene splicing, have provided humanity with tools to manipulate DNA with unprecedented precision. These advancements open the door to exploring the potential for creating hybrid organisms, both terrestrial and extraterrestrial. Theoretical frameworks suggest that if alien DNA were introduced into human genetic material, the resulting hybrids could possess enhanced abilities or traits, drawing parallels to myths and folklore of beings with extraordinary powers. This scientific foundation raises the question of how such hybrids might be perceived within the broader context of evolutionary biology and what that means for the future of humanity.

The psychological effects of hybridization extend beyond mere genetic implications. The concept of being part alien carries profound identity challenges and existential questions for individuals who may feel a sense of otherness. This psychological burden can manifest in various ways, influencing personal relationships and societal interactions. The belief in one's hybrid status can lead to a profound sense of isolation or, conversely, an inflated sense of purpose. Consequently, the mental health implications of these beliefs are critical to understanding the broader cultural impacts of hybridization narratives.

Ethical considerations surrounding human-alien hybridization are fraught with ambiguity. The potential for exploitation, informed consent, and the commodification of life forms poses significant moral dilemmas. As humanity stands on the precipice of genetic manipulation, the question arises: should we pursue the creation of hybrids, and at what cost? The intersection of ethics, science, and the spiritual implications of hybrid existence invites a dialogue that transcends mere speculation, challenging our fundamental beliefs about life, consciousness, and the cosmos.

Cultural interpretations of hybrid beings have permeated literature and film, reflecting societal anxieties and aspirations regarding the unknown. Sci-fi narratives often serve as allegories for the human experience, contemplating themes of identity, belonging, and the fear of the other. Historical accounts of alien encounters further complicate the discourse, as they bring to light a rich tapestry of belief systems and experiences that shape public perception of hybrids. As these narratives evolve, they will continue to influence how society perceives the implications of genetic alchemy and the potential realities of hybrid beings, ultimately shaping our understanding of what it means to be human in an increasingly complex universe.

### CASE STUDIES OF ALLEGED HYBRIDS

The phenomenon of alleged human-alien hybrids has intrigued researchers and enthusiasts alike, giving rise to numerous case studies that explore the supposed intersections of human genetics and extraterrestrial influence. One of the most widely discussed cases is that of the so-called "Star Children," a term used to describe individuals believed to possess unique genetic traits attributed to alien ancestry. These individuals often report heightened psychic abilities, unusual physical characteristics, and a profound sense of being different from their peers. Their experiences frequently include memories of abduction and encounters with extraterrestrial beings, leading some to theorize that these hybrids represent a new stage in human evolution, inherently linked to advanced alien civilizations.

Another compelling case is that of the alleged hybridization program conducted by the Greys, a type of extraterrestrial commonly associated with abduction narratives. Numerous individuals claim to have undergone procedures in which their genetic material was extracted and manipulated, resulting in the creation of hybrid offspring. These hybrids are often described as possessing a blend of human and alien features, and many witnesses assert that they have encountered these beings in various contexts, from domestic environments to clandestine government facilities. Such reports raise questions regarding the ethical implications of genetic manipulation and the potential consequences of integrating extraterrestrial DNA into the human genetic pool.

The psychological effects of hybridization are another critical aspect of the discourse surrounding these case studies. Individuals who identify as hybrids or who believe they have been involved in hybridization often experience significant mental and emotional distress. They may grapple with feelings of isolation, confusion, and a disconnection from mainstream society. Furthermore, the stigma associated with hybrid claims can exacerbate these psychological struggles, leading to a growing community of individuals seeking validation and support. This phenomenon highlights the need for a nuanced understanding of the psychological ramifications of alleged alien encounters and the potential for therapeutic interventions that address these unique experiences.

Culturally, the narratives surrounding hybrids have permeated various forms of art and literature, reflecting society's fascination with the concept of genetic alchemy. Sci-fi literature and film often depict hybrid beings as symbols of humanity's fears and aspirations regarding the unknown. From the classic works of Philip K. Dick to contemporary series that explore themes of identity and belonging, the portrayal of hybrids challenges conventional notions of what it means to be human. This cultural lens not only serves as a commentary on our technological advancements but also invites a broader discussion on the implications of altering human genetics, both in fiction and potential reality.

Lastly, government conspiracy theories surrounding hybridization add another layer of complexity to the discourse. Many theorists assert that covert governmental programs are actively researching hybrid beings, suggesting that the existence of hybrids is not merely the result of extraterrestrial encounters but also a product of human experimentation. This perspective invokes a distrust in governmental transparency and raises ethical questions about the role of authority in matters of genetic engineering. As such, the case studies of alleged hybrids serve not only to illuminate individual experiences but also to underscore a larger societal dialogue about the moral and ethical implications of humanity's relationship with alien life and the potential for hybrid beings to redefine our understanding of existence itself.

## ~ 3 ~

# GENETIC ENGINEERING AND MANIPULATION

### TECHNIQUES IN GENETIC MODIFICATION

Techniques in genetic modification have evolved significantly over the years, blending the boundaries between science fiction and reality. Among UFO enthusiasts and conspiracy theorists, the discourse surrounding these techniques often delves into the realm of alien engineering. Current methodologies such as CRISPR-Cas9 offer unprecedented precision in gene editing, enabling scientists to alter specific genes with remarkable accuracy. This technology raises intriguing questions about the potential for human-alien hybridization, as it could theoretically allow for the incorporation of extraterrestrial DNA into human genomes. Such possibilities spark the imagination and fuel theories about the existence of hybrid beings living among us.

In the context of genetic manipulation, understanding the psychological effects of hybridization is paramount. The prospect of human-alien hybrids challenges our understanding of identity and consciousness. Individuals who believe they have experienced contact with extraterrestrial beings often report profound psychological changes, including altered perceptions of self and reality. These experiences can lead to feelings of alienation or enlightenment, depending on the individual's perspective. The implications of such psychological transformations are vast, influencing not only personal beliefs but also broader cultural narratives surrounding hybrid beings.

Ethical considerations are equally critical in discussions of genetic modification, especially regarding alien genetics. As techniques advance, the line between ethical and unethical practices becomes increasingly blurred. Debates surrounding the morality of altering human DNA, particularly with the intent of creating hybrids, raise questions about consent, the definition of humanity, and the potential for exploitation. Conspiracy theories often emerge in this context, suggesting that government entities may engage in secretive programs to manipulate human genetics for unknown purposes. This fosters a climate of distrust and suspicion among those who suspect that hybridization experiments are taking place without public knowledge.

Cultural interpretations of hybrid beings have also played a significant role in shaping public perceptions of genetic modification. From ancient mythologies to modern science fiction literature and film, the concept of hybrids has been a source of fascination and fear. These narratives often reflect societal anxieties about the unknown and the implications of crossing natural boundaries. They serve as a mirror to our own ethical dilemmas, depicting hybrids as either saviors or harbingers of doom. As these stories continue to proliferate, they influence how individuals approach the reality of genetic engineering and the potential for human-alien connections.

Finally, examining historical accounts of alien encounters provides a rich tapestry of evidence that some believe supports the idea of hybridization. Reports from credible witnesses often describe beings with features that defy conventional human categorization, suggesting the presence of a hybrid lineage. These encounters, whether interpreted as genuine experiences or psychological phenomena, contribute to the ongoing discourse about our place in the universe and the potential for interspecies relationships. As medical research on hybrid physiology develops, it may unveil new dimensions of understanding about the limits and possibilities of human genetics, pushing the boundaries of what it means to be human in an ever-expanding cosmic landscape.

## THE ROLE OF CRISPR AND OTHER TECHNOLOGIES

The advent of CRISPR technology has revolutionized the landscape of genetic engineering, providing unprecedented tools for precise editing of DNA sequences. This technology, which stands for Clustered Regularly Interspaced Short Palindromic Repeats, allows scientists to modify genetic material with a level of accuracy that was previously unattainable. For UFO enthusiasts and conspiracy theorists, the implications of CRISPR extend far beyond terrestrial applications. The potential for this technology to be employed in the creation of human-alien hybrids raises profound questions about the manipulation of genetics and the ethical boundaries that may be crossed in the pursuit of such endeavors.

The intersection of CRISPR and other genetic engineering technologies invites speculation about the possible existence of advanced extraterrestrial civilizations capable of similar or superior genetic manipulations. If alien beings possess the knowledge and tools to alter genetic makeup, the implications for hybridization become staggering. Theories abound regarding whether such technologies could have already been introduced to humanity, either through direct contact with extraterrestrial intelligences or through reverse engineering of alien technology. This premise fuels the belief that humans may unknowingly be participating in an ongoing experiment of hybridization, where CRISPR-like techniques are utilized to create beings that blur the lines between species.

Moreover, the psychological effects of hybridization are a critical area of exploration. As advancements in genetic engineering unfold, they raise questions about identity, consciousness, and the very essence of what it means to be human. Individuals who believe they are hybrids or descendants of hybrid beings may experience unique psychological phenomena, including a sense of disconnection from their own humanity or a feeling of being part of a larger cosmic narrative. The potential for CRISPR to enhance or alter human traits could deepen these experiences, leading to an existential crisis for those grappling with their perceived role in a universe that may harbor other intelligent life forms.

The ethical implications of manipulating alien genetics using technologies like CRISPR cannot be overstated. As humanity stands on the brink of unprecedented genetic capabilities, the moral responsibilities of such power become increasingly complex. Questions arise regarding consent, the natural order, and the potential for creating beings that may suffer or possess cognitive abilities beyond human comprehension. For conspiracy theorists, the existence of government programs that may be clandestinely experimenting with alien genetics further complicates the ethical landscape. The fear that these programs operate without public oversight or ethical scrutiny may fuel beliefs in covert agendas aimed at creating a new class of hybrid beings.

Culturally, the interpretations of hybrid beings within sci-fi literature and film have often reflected societal anxieties about genetic manipulation and alien contact. These narratives provide a lens through which to examine our fears and hopes regarding the merging of human and alien traits. Historical accounts of alien encounters frequently include descriptions of beings that exhibit hybrid characteristics, suggesting a long-standing fascination with the idea of genetic fusion. As genetic technologies evolve, so too do our cultural interpretations of what it means to be a hybrid. The spiritual perspectives on hybrid existence also offer rich ground for exploration, as they challenge conventional views of life and consciousness, inviting us to consider the possibility of an interconnected cosmic existence that transcends traditional human boundaries.

## ~ 4 ~

# PSYCHOLOGICAL EFFECTS OF HYBRIDIZATION

### IDENTITY CRISIS IN HYBRIDS

Identity crises in hybrids emerge as a complex intersection of genetics, psychology, and cultural narratives. Individuals who find themselves straddling the line between human and alien often grapple with profound questions of self-worth, belonging, and the very essence of what it means to be human. The hybrid experience can lead to feelings of alienation, as they are seen as anomalies in both human and extraterrestrial contexts. This dissonance breeds a unique psychological landscape where hybrids must navigate their dual identities, often resulting in internal conflicts that can manifest as anxiety, depression, or a deep sense of isolation.

From a psychological standpoint, the implications of hybridization extend beyond mere identity confusion. The merging of human and alien DNA introduces not only physical changes but also cognitive and emotional ones. Hybrids may possess enhanced abilities or heightened sensitivities, which can further alienate them from their human counterparts. This enhanced perception can lead to a greater awareness of societal injustices or existential dilemmas, potentially propelling hybrids into roles as advocates or healers. However, this heightened awareness can also be a double-edged sword, creating a chasm between their experiences and those of ordinary humans, exacerbating their identity crisis.

Culturally, hybrids occupy a unique space in the collective imagination, often depicted in literature and film as either saviors or harbingers of doom. These portrayals influence public perception and can contribute to the stigma or reverence surrounding hybrid beings. Sci-fi narratives frequently explore the tension between the hybrid's human heritage and alien origins, reflecting societal fears and hopes regarding the unknown. Such cultural interpretations can either validate the hybrid's existence and struggles or dehumanize them further, complicating their sense of identity. The narratives woven into popular culture thus play a significant role in shaping how hybrids perceive themselves and how they are perceived by others.

Ethically, the existence of hybrids raises numerous questions about consent, agency, and the moral implications of genetic experimentation. The notion of human-alien hybridization challenges our understanding of identity and the rights of beings who may not fit neatly into established categories. As we delve deeper into the implications of alien engineering, we must confront the ethical dilemmas surrounding the creation of hybrids and the responsibilities of those who engage in such manipulations. This ethical landscape demands a reevaluation of what it means to be human and how we define personhood, particularly for those who embody traits from multiple species.

Finally, the spiritual dimensions of hybrid existence suggest that these beings may serve as bridges between worlds, offering insights into the nature of consciousness and the universe. Many hybrids report experiences that transcend traditional understandings of reality, engaging with spiritual practices that reflect their unique heritage. This spiritual perspective can foster a sense of purpose and belonging, even amidst the chaos of their identity crisis. As we explore the multifaceted identity crises faced by hybrids, we uncover not only the struggles inherent in their existence but also the potential for profound growth and transformation that lies in their journey of self-discovery.

## PSYCHOLOGICAL IMPACT ON HUMAN PARENTS

The psychological impact on human parents of alleged alien hybrids is a multifaceted issue that encompasses emotional, cognitive, and social dimensions. Many individuals who believe they have been involved in hybridization experiences report profound psychological effects, ranging from anxiety and confusion to a sense of alienation from their own communities. These parents often grapple with the implications of their experiences, which can lead to a reevaluation of their beliefs about family, identity, and even the nature of humanity itself. The struggle to reconcile these profound experiences with everyday life can result in significant mental distress, compelling parents to seek support in niche communities that often validate their experiences.

Parents of hybrid children frequently experience intense feelings of isolation, as societal norms and expectations do not typically accommodate the existence of extraterrestrial involvement in human reproduction. This alienation can manifest in various ways, including difficulties in forging connections with family and friends who may dismiss or ridicule their claims. The stigma associated with being a so-called 'hybrid parent' can exacerbate feelings of loneliness and despair, making it challenging for these individuals to navigate their emotional landscape. In this context, support networks, often found within UFO enthusiast circles or online forums, become crucial for sharing experiences and coping strategies.

Moreover, the cognitive dissonance experienced by these parents can lead to an ongoing internal conflict regarding their sense of reality. Many report a struggle between the desire to accept their experiences as genuine and the pressure to conform to mainstream societal views. This conflict can result in a fragmented sense of self, where the individual is torn between two opposing narratives: one rooted in conventional human experience and the other suggesting a radical redefinition of human existence. Such mental turmoil often necessitates the exploration of alternative belief systems and spiritual frameworks as individuals seek meaning and understanding in their altered reality.

The ethical implications of hybridization further complicate the psychological landscape for these parents. The moral dilemmas surrounding consent, parental rights, and the potential exploitation of hybrid beings can weigh heavily on those who feel a connection to these entities. Parents may find themselves questioning the very nature of their role as caregivers and advocates for their children, particularly when faced with the uncertainty of their children's identities and purposes. This ethical quandary can lead to existential crises, where individuals must confront uncomfortable truths about autonomy, agency, and the implications of human-alien relationships.

Cultural interpretations of hybrid beings also play a significant role in shaping the psychological experiences of human parents. Narratives surrounding alien encounters—whether in folklore, literature, or film—often influence how individuals perceive their own situations. These cultural artifacts can both inspire and terrify, providing frameworks through which parents conceptualize their experiences. The blending of myth and reality in these narratives can lead to further psychological conflict, as parents navigate their own unique stories while grappling with the pervasive influence of societal myths regarding alien life. Ultimately, the psychological impact on human parents of alleged alien hybrids is a complex interplay of emotions, beliefs, and cultural narratives, reflecting broader themes of identity and existence in a universe that may be far more intricate than previously understood.

# ~ 5 ~

# ETHICAL IMPLICATIONS OF ALIEN GENETICS

## MORALITY OF GENETIC MANIPULATION

The morality of genetic manipulation, particularly in the context of human-alien hybridization, raises profound ethical questions that resonate deeply within the realms of UFO enthusiasts and conspiracy theorists. As we delve into this complex subject, we must acknowledge the dual-edged nature of genetic engineering. On one hand, it holds the potential for groundbreaking advancements in medicine and agriculture; on the other, it presents significant moral dilemmas. The prospect of creating beings that blur the lines between human and extraterrestrial invokes fears of dehumanization and exploitation, challenging our understanding of what it means to be sentient.

The implications of altering human genetics through alien technology extend beyond mere scientific inquiry; they touch upon fundamental human rights and the sanctity of life. If we consider the ethical ramifications of creating hybrids, we must grapple with questions of consent, identity, and autonomy. Are these beings entitled to rights and protections, or do they exist in a legal and moral gray area? These concerns are compounded by historical accounts of alien encounters, where the narratives often depict abductions and experimentation, casting a shadow of distrust over the intentions of those who might engage in genetic manipulation.

Cultural interpretations of hybrid beings further complicate the moral landscape. Throughout history, hybrids have been portrayed in various ways, from revered demigods in mythology to monstrous abominations in modern horror. These portrayals influence public perception and acceptance of genetic manipulation, shaping the moral discourse surrounding it. As society grapples with the reality of potential hybridization, these cultural narratives can either serve as cautionary tales or as frameworks for understanding the complexities of hybrid existence.

In the realm of psychological effects, the implications of hybridization extend into the psyche of both the hybrids themselves and the societies that produce them. The alienation experienced by hybrids—who may struggle with their identity and sense of belonging—presents a unique set of challenges. This psychological dimension raises ethical questions about the responsibilities of those who engage in genetic manipulation. Are we prepared to support and integrate these beings into our societies, or will they become scapegoats for our fears and prejudices?

Finally, the intersection of government conspiracy theories and medical research on hybrid physiology highlights the need for transparency and ethical oversight in genetic manipulation. As clandestine programs and covert experiments are rumored to exist, the potential for abuse becomes a pressing concern. The moral framework guiding genetic research must prioritize the welfare of all sentient beings, ensuring that the pursuit of knowledge does not come at the expense of ethical integrity. As we navigate this uncharted territory, the discourse surrounding the morality of genetic manipulation will continue to evolve, reflecting our deepest fears, hopes, and aspirations for the future.

## LEGAL CONSIDERATIONS SURROUNDING HYBRIDS

Legal considerations surrounding hybrids, particularly in the context of human-alien hybridization, present a complex and often contentious landscape. As advancements in genetic engineering prompt discussions about the ethical and legal status of hybrids, fundamental questions regarding personhood, rights, and the implications of their existence arise. Current legal frameworks are ill-equipped to address the nuances of hybrids resulting from extraterrestrial interaction, leaving room for interpretation and debate among lawmakers, ethicists, and the public.

In many jurisdictions, the legal definitions of human and non-human entities are strictly delineated. Hybrids, which may possess both human and non-human DNA, challenge these definitions, complicating their status under existing laws. Legal scholars advocate for a reevaluation of personhood that includes hybrids, suggesting that as these beings potentially possess cognitive and emotional capacities similar to humans, they should be granted certain rights and protections. This debate echoes broader discussions surrounding the rights of animals and artificial intelligences, as society grapples with the implications of creating life forms that do not fit neatly into established categories.

The potential for hybrids to be used in medical research raises further legal and ethical concerns. If hybrids are viewed as a new species or a distinct entity, regulations governing biomedical research must be adapted to include them. Researchers could face legal challenges related to consent, ownership of genetic material, and the treatment of hybrids in clinical settings. Moreover, the question of whether hybrids can legally consent to participate in research is particularly fraught, especially given the uncertain understanding of their cognitive abilities and emotional awareness.

Cultural interpretations of hybrids also play a crucial role in shaping legal frameworks. In many cultures, hybrids are depicted as either saviors or threats, influencing public perception and, consequently, legislative action. This cultural lens can lead to stigmatization or protection, depending on prevailing narratives about hybrids in literature, film, and folklore. As these narratives evolve, they may impact legal definitions and the social acceptance of hybrids, further complicating the legal landscape.

Finally, government conspiracy theories surrounding hybrids often intertwine with legal considerations. The existence of clandestine programs allegedly focused on hybridization raises questions about governmental accountability and transparency. If hybrids are indeed being created and studied in secret, the legal implications regarding human rights violations, ethical breaches in scientific research, and the potential for exploitation become paramount. As society seeks to understand and regulate hybrid beings, a comprehensive legal framework that addresses these multifaceted issues becomes increasingly necessary, ensuring that the rights and dignity of all forms of life are respected in this uncharted territory.

## ~ 6 ~

# CULTURAL INTERPRETATIONS OF HYBRID BEINGS

MYTHOLOGICAL PERSPECTIVES

Mythological perspectives on alien hybridization reveal fascinating parallels between ancient narratives and contemporary experiences reported by individuals who believe they are hybrids. Throughout history, various cultures have depicted beings that share characteristics with both humans and divine or extraterrestrial entities. These mythological figures often possess extraordinary abilities, which can be interpreted as reflections of the traits attributed to modern-day hybrids. The blending of human and otherworldly traits in mythology serves as a precursor to today's discussions surrounding genetic manipulation and hybridization, suggesting an enduring fascination with the idea of transcending human limitations through external influences.

The psychological effects of hybridization, as depicted in mythological tales, offer insights into the human psyche's response to encountering the unknown. Characters such as the Nephilim from biblical texts or the demigods of Greek mythology often experience internal conflicts regarding their identities, straddling the line between mortal and divine. These narratives resonate with contemporary accounts of individuals who describe feelings of alienation, identity confusion, and a sense of being "different" from their peers. Such psychological ramifications underscore the potential emotional and cognitive consequences of hybridization, both in ancient lore and modern experiences, signifying a persistent struggle with self-identity when faced with extraordinary circumstances.

Moreover, cultural interpretations of hybrid beings across various mythologies highlight the ethical implications of alien genetics. Many stories reflect societal fears and aspirations regarding the manipulation of life forms, raising questions about the morality of altering human traits through alien intervention. This ethical discourse is mirrored in modern debates surrounding genetic engineering, where the power to change human biology brings forth concerns about eugenics, consent, and the potential creation of a new class of beings. By examining mythological perspectives, we gain a deeper understanding of the moral dilemmas that may arise when humanity encounters the possibility of hybrid existence.

In the realm of science fiction literature and film, mythological themes of hybridization continue to influence narratives that captivate audiences. Stories that explore the intricacies of human-alien relationships often draw on archetypal figures from mythology, using them to frame modern anxieties about technology, evolution, and the potential for new life forms. These narratives not only entertain but also provoke critical thought about the implications of hybridization, emphasizing the timeless nature of these themes across cultures and eras. As such, mythological perspectives enrich the discourse on hybridization by providing a narrative framework through which we can examine our hopes and fears regarding the future of humanity.

Historical accounts of alien encounters, intertwined with mythological elements, further shape our understanding of hybridization. Many ancient civilizations recorded experiences that can be interpreted as interactions with otherworldly beings, often describing them in terms reminiscent of mythological stories. These accounts contribute to the ongoing debate about the reality of hybridization, suggesting that the fascination with alien beings may not solely be a modern phenomenon but rather a continuation of humanity's quest to understand its place in the cosmos. By exploring these historical narratives, we can better appreciate how mythological perspectives inform our contemporary views on alien engineering, genetic manipulation, and the very essence of what it means to be human.

## MODERN CULTURAL REPRESENTATIONS

Modern cultural representations of alien hybridization are pervasive, permeating various forms of media, including literature, film, and visual arts. In recent decades, these representations have shifted from the sensationalized depictions of extraterrestrial beings to more nuanced portrayals that explore the complexities of identity, belonging, and the human experience. This evolution reflects a growing societal intrigue with the concept of hybrid beings, prompting audiences to engage with the philosophical and ethical dilemmas inherent in the idea of human-alien intermingling. The narratives surrounding these hybrids often serve as a mirror to contemporary anxieties about genetics, technology, and the future of humanity.

The intersection of genetic engineering and hybridization is a recurring theme in modern storytelling. Works of science fiction frequently grapple with the implications of manipulating DNA, presenting scenarios where humans and aliens merge in ways that challenge traditional notions of what it means to be human. These narratives often explore the psychological effects of such hybridization, delving into themes of alienation, identity crisis, and the quest for acceptance. By examining the inner lives of hybrids, creators offer insights into the human condition, prompting audiences to reflect on their own experiences with difference and the search for belonging in an increasingly complex world.

Cultural interpretations of hybrid beings also raise significant ethical questions. As society navigates the rapid advancements in genetic technology, concerns regarding consent, autonomy, and the potential for exploitation emerge. Modern representations often highlight these ethical dilemmas, urging viewers to critically consider the moral implications of playing God through genetic manipulation. This discourse is particularly relevant in light of ongoing debates surrounding bioethics and the regulation of genetic research. The portrayal of hybrids in media serves as a catalyst for discussion, pushing audiences to confront uncomfortable truths about the direction of scientific progress and its impact on the fabric of society.

Moreover, historical accounts of alien encounters have shaped contemporary cultural representations. Tales of abduction and clandestine government experiments fuel conspiracy theories that intertwine with the narrative of hybridization. These accounts often reflect societal fears and fascinations with the unknown, blending fact with fiction in a way that captivates the imagination. As these historical narratives evolve, they inform modern interpretations of hybrids, suggesting that the boundaries between myth and reality are increasingly porous. This interplay invites audiences to question the authenticity of their beliefs and the sources of their information in a world rife with misinformation.

Finally, the spiritual perspectives on hybrid existence offer another layer of complexity to modern cultural representations. Many communities view hybrids not merely as products of genetic manipulation but as beings with a unique purpose in the cosmic landscape. These spiritual interpretations often posit that hybrids embody a bridge between humanity and the extraterrestrial, serving as conduits for higher knowledge and understanding. Such views challenge conventional paradigms of evolution and existence, inviting deeper exploration into the nature of consciousness and the potential for spiritual evolution through hybridization. As cultural representations continue to evolve, they reflect an ongoing fascination with the possibilities of genetic alchemy and the profound questions it raises about our place in the universe.

## ~ 7 ~

# HYBRIDIZATION IN SCI-FI LITERATURE AND FILM

NOTABLE WORKS FEATURING HYBRIDS

Notable works featuring hybrids have permeated various forms of media, illustrating the complex interplay between human and alien genetics. In literature, the notion of hybrid beings often reflects societal anxieties about the unknown and the manipulation of what it means to be human. Classic novels such as "The Midwich Cuckoos" by John Wyndham explore the psychological implications of alien hybridization, portraying a village where women mysteriously conceive children with extraordinary abilities. The narrative evokes fear and fascination, prompting readers to consider the ethical ramifications of such genetic interventions.

Film has also played a crucial role in shaping perceptions of hybrid beings. Movies like "Species" and "Annihilation" delve into the consequences of human-alien interaction, often highlighting the unpredictable nature of hybrid creatures. These films address the potential for both beauty and monstrosity in hybrids, reflecting cultural interpretations of alien genetics as a double-edged sword. The visceral portrayal of hybrids in visual media serves to amplify the psychological impact on audiences, often leaving them to grapple with their own beliefs about identity and the boundaries of humanity.

In the realm of conspiracy theories, notable accounts of alleged hybridization have emerged, often intertwined with government secrecy and military experimentation. Reports from individuals claiming to have undergone abduction experiences frequently include depictions of hybrid offspring, raising questions about the ethics of such practices. These narratives suggest a hidden agenda behind advanced genetic manipulation, feeding into broader fears of government surveillance and control. The intersection of personal testimony and speculative fiction creates a fertile ground for exploring the darker aspects of hybridization, pushing the boundaries of credibility and belief.

Scientific discourse surrounding hybrid beings has also influenced public perception, particularly in the fields of evolutionary biology and medical research. Scholars examine the potential physiological differences and adaptations that hybrids could exhibit, fostering a dialogue between science fiction and real-world genetic engineering. This academic inquiry not only legitimizes the concept of hybrids but also challenges existing ethical frameworks regarding genetic modification. The implications of creating hybrid entities extend beyond the laboratory, prompting discussions about the moral responsibilities of scientists and the societal impacts of their innovations.

Finally, spiritual perspectives on hybrid existence reveal a deeper layer of cultural interpretation surrounding the phenomenon. Beliefs in soul integration and the spiritual evolution of hybrids often reflect humanity's quest for understanding its place in the cosmos. These views posit that hybrids may possess unique insights or abilities that bridge the gap between human and extraterrestrial experiences. By integrating spiritual narratives with the scientific and fictional elements of hybridization, a more comprehensive understanding emerges, inviting UFO enthusiasts and conspiracy theorists alike to ponder the mysteries that lie at the intersection of humanity and the unknown.

### ANALYSIS OF THEMES AND MOTIFS

The analysis of themes and motifs in "Genetic Alchemy: The Dark Art of Alien Engineering" reveals a complex interplay of human fascination, fear, and ethical considerations surrounding the concept of alien hybridization. Central to this narrative is the theme of identity, as the existence of hybrid beings challenges traditional notions of what it means to be human. This exploration of identity not only engages with the biological aspects of hybridization but also delves into the psychological ramifications for individuals who may grapple with their sense of self in a world where the boundaries between species are increasingly blurred. The book invites readers to consider how these hybrid beings could embody both a threat and a promise, reflecting humanity's deepest anxieties and aspirations.

Another significant motif in the text is the juxtaposition of science and mysticism. The processes of genetic engineering and manipulation are often portrayed in stark, clinical terms, yet they are juxtaposed with spiritual interpretations of hybrid existence. This duality raises questions about the ethical implications of such scientific endeavors. Are we playing god by tampering with the genetic fabric of life, or are we merely uncovering the secrets of our own existence? This motif invites UFO enthusiasts and conspiracy theorists alike to contemplate the potential consequences of hybridization not just on a biological level, but also in terms of spiritual evolution and our understanding of consciousness.

Cultural interpretations of hybrid beings emerge as another essential theme, illustrating how different societies perceive and react to the idea of alien-human hybrids. The book highlights historical accounts of alien encounters, showcasing how these narratives have evolved over time, reflecting societal fears, desires, and moral quandaries. The portrayal of hybrids in various cultures—ranging from revered figures to monstrous beings—underscores the significant role that myth and folklore play in shaping contemporary beliefs about extraterrestrial life. This exploration encourages readers to consider how cultural contexts influence the reception of hybridization theories and the broader implications for humanity's place in the cosmos.

The psychological effects of hybridization are also a crucial theme, particularly in understanding the emotional and mental health challenges faced by individuals who believe they are hybrids or have undergone alien encounters. The narrative delves into the complexities of trauma, identity crisis, and the search for belonging among those who feel caught between two worlds. This theme resonates deeply within the niche of psychological research, as it raises pertinent questions about the impact of perceived alien influence on human psychology. It invites readers to explore the blurred lines between reality and delusion, as well as the potential for healing and understanding within the hybrid experience.

Finally, the ethical implications of alien genetics are examined through the lens of governmental conspiracy theories surrounding hybrid programs. The book scrutinizes the tension between scientific advancement and moral responsibility, questioning who holds authority in making decisions about genetic manipulation. This theme resonates with those concerned about the potential for misuse of scientific knowledge and the erosion of ethical boundaries. By addressing the intersection of power, knowledge, and hybridization, "Genetic Alchemy" compels readers to confront the moral dilemmas that arise when humanity stands at the crossroads of evolution and extraterrestrial influence, leaving them to ponder the future trajectory of our species in the face of such profound possibilities.

## ~ 8 ~

# HISTORICAL ACCOUNTS OF ALIEN ENCOUNTERS

ANCIENT TEXTS AND ARTIFACTS

Ancient texts and artifacts have long fascinated those who explore the intersection of humanity and extraterrestrial influences. Historical documents, ranging from the Sumerian tablets to the biblical accounts, often contain references that suggest interactions with beings not of this world. These texts frequently describe gods and celestial visitors who impart knowledge or genetic gifts to humanity, sparking debate over whether these entities might have been advanced extraterrestrials. For UFO enthusiasts and conspiracy theorists, these narratives serve as tantalizing evidence of alien contact and the potential for human-alien hybridization throughout history.

The study of ancient artifacts, such as the Nazca Lines in Peru or the intricate stone carvings of the Mayans, raises questions about the technological prowess of early civilizations. Many believe that the precision and complexity of these creations could not have been achieved without outside guidance. Theories suggest that these societies may have received assistance from alien visitors, leading to advancements in agriculture, architecture, and even governance. Such speculations often feed into the broader narrative of alien genetic manipulation, where the idea posits that extraterrestrial beings not only visited Earth but also played a role in shaping human evolution.

Cultural interpretations of hybrid beings across various civilizations provide a rich tapestry of insights into humanity's perception of its own origins. In many ancient cultures, hybrid figures like the demigods of Greek mythology or the Anunnaki of Mesopotamian lore represent a bridge between the divine and the mortal. These stories often reflect a societal understanding of genetic intermingling, whether literal or metaphorical, inviting further inquiry into how these ancient narratives parallel contemporary discussions about alien hybridization and genetic engineering. The psychological effects of these beliefs can be profound, influencing how individuals perceive their identity and place within the universe.

The ethical implications surrounding the notion of alien genetics raise significant questions. If ancient texts and artifacts suggest a history of hybridization, what moral responsibilities accompany such knowledge? As modern science explores genetic manipulation, parallels can be drawn to the ancient practices of altering crops or breeding animals, yet the stakes are higher when considering human genetics. The potential for creating hybrid beings in a laboratory setting, inspired by historical accounts of extraterrestrial encounters, prompts urgent discussions about consent, identity, and the essence of humanity itself.

In the realm of science fiction, the themes of hybridization and alien influence are ever-present, reflecting societal anxieties and hopes regarding our future. Works of fiction often extrapolate on the implications of genetic engineering and hybrid beings, exploring both utopian and dystopian outcomes. These narratives resonate with historical accounts of alien encounters, enriching the conversation around human evolution and the possible paths our species may take. As the boundaries between science, myth, and ethics continue to blur, the exploration of ancient texts and artifacts becomes not only an academic pursuit but also a lens through which we can better understand our own existence and the potential legacy of hybridization.

## MODERN-DAY SIGHTINGS AND REPORTS

The contemporary landscape of UFO sightings and reports has expanded dramatically, fueled by advancements in technology and a growing cultural acceptance of the extraterrestrial narrative. In recent years, numerous accounts have emerged from individuals claiming direct encounters with unidentified aerial phenomena (UAP) and alleged alien beings. These reports often include intricate details that suggest not only visual anomalies but also experiences of hybridization or interactions with entities believed to possess advanced genetic manipulation capabilities. As the dialogue around such phenomena grows, the implications of these encounters extend into the realms of genetics and the psychological impacts on those involved.

Witnesses frequently describe scenarios that evoke a sense of disorientation, often accompanied by missing time and vivid recollections of contact with beings who exhibit not only human-like characteristics but also distinct differences. These descriptions resonate with historical accounts of hybrid beings, suggesting a continuity in the narratives that span decades, if not centuries. The psychological effects reported by these individuals are profound, often leading to feelings of isolation, anxiety, and a questioning of their very identity. Such experiences raise critical questions about the psychological ramifications of hybridization and the potential for deeper genetic ties that may exist between humans and these entities.

Moreover, the ethical implications of alien genetics have sparked significant debate among enthusiasts and researchers alike. As reports of hybridization continue to surface, the moral considerations surrounding consent, genetic manipulation, and the potential for creating new life forms come into focus. The ethical discourse is further complicated by government involvement and the secrecy surrounding UFO phenomena, leading many to speculate about the existence of clandestine programs aimed at exploring the possibilities of human-alien hybridization. This secrecy has fostered a fertile ground for conspiracy theories, as individuals seek to reconcile their experiences with the lack of transparency from authorities.

Cultural interpretations of hybrid beings have also evolved, influenced by literature and film that explore the intersection of humanity and alien life. Sci-fi narratives frequently delve into the complexities of coexistence and the potential for hybridization, reflecting societal fears and aspirations. These cultural artifacts shape public perception, providing a framework through which individuals can understand and contextualize their experiences. The representation of hybrids in popular media often blurs the line between fiction and reality, leading to an increased fascination with the possibilities of genetic alchemy and its implications for humanity's future.

As interest in the phenomenon persists, the academic study of hybridization and its implications for evolutionary biology becomes increasingly relevant. Researchers are examining the potential for hybrid vigor, adaptability, and the long-term consequences of interspecies genetic exchange. Additionally, spiritual perspectives on hybrid existence offer an intriguing angle, suggesting that these experiences may lead to a deeper understanding of consciousness and existence beyond traditional boundaries. As modern-day sightings and reports continue to emerge, they serve as a catalyst for ongoing inquiry into the profound questions surrounding our place in the universe and the enigmatic nature of life itself.

# ~ 9 ~

# HYBRIDIZATION AND EVOLUTIONARY BIOLOGY

THEORIES OF EVOLUTION AND HYBRIDIZATION

The concept of evolution has long fascinated both scientists and the general public, especially when intertwined with the idea of hybridization. Theories of evolution traditionally emphasize natural selection, genetic drift, and mutation as primary mechanisms driving the development of species. However, when one considers the potential influence of extraterrestrial beings, the framework of evolution expands dramatically. The introduction of alien genetics into human DNA raises questions about the very mechanisms of evolution itself, suggesting that hybridization could serve as a catalyst for rapid evolutionary changes, potentially leading to new human capabilities or even entirely new subspecies.

Hybridization, in this context, refers not only to the mixing of human and alien genetic material but also to the broader implications of such interactions. The practice of genetic engineering has already shown us that manipulation of DNA can yield astonishing results, from disease resistance to enhanced physical traits. When examining potential human-alien hybrids, one must consider the psychological impacts of this genetic interplay. The merging of two distinct consciousnesses could lead to unique cognitive abilities, altered perceptions of reality, and perhaps even a new understanding of existence itself. These psychological effects could redefine what it means to be human, challenging our current definitions of identity and consciousness.

Ethical considerations surrounding alien genetics are paramount in discussions of hybridization. The implications of creating hybrids raise profound questions regarding autonomy, consent, and the moral responsibilities of those engaged in such practices. Are hybrids entitled to the same rights as pure humans? How would society integrate beings that embody characteristics from both human and alien lineages? These inquiries not only touch upon the legal and ethical frameworks of our world but also evoke deep cultural interpretations of what it means to be "other." The fear of the unknown, often depicted in science fiction literature and film, can manifest in societal backlash against these hybrids, echoing historical prejudices against those who deviate from the norm.

Cultural narratives surrounding hybrid beings have been prevalent throughout history, often reflecting societal anxieties and hopes. From ancient mythologies to modern sci-fi tales, the hybridization motif serves as a mirror for our deepest fears and aspirations regarding evolution and otherness. The portrayal of hybrids in literature and film often oscillates between the monstrous and the sublime, influencing public perception and shaping the discourse on alien interactions. These artistic interpretations can either reinforce skepticism about government conspiracies related to alien experimentation or inspire intrigue in the potential of hybrid beings to transform humanity's future.

In the realm of scientific inquiry, the study of hybrid physiology opens new avenues for medical research. Understanding the biological makeup of hybrids could lead to breakthroughs in genetics, offering insights into human health and disease. However, this scientific exploration is inevitably intertwined with conspiracy theories that suggest government agencies have long engaged in clandestine research on hybrids. Unraveling the complexities of hybridization involves not only a rigorous examination of evolutionary biology but also an exploration of the spiritual dimensions of existence and the broader implications of sharing our world with beings of potentially superior intelligence. The intersection of these themes presents a rich tapestry for further exploration, inviting both skepticism and wonder about the future of humanity in an increasingly complex universe.

## IMPLICATIONS FOR HUMAN EVOLUTION

The implications of human evolution in the context of genetic manipulation and alien hybridization present a complex tapestry of possibilities that challenge our understanding of humanity's origins and future. As we delve into the narrative of genetic alchemy, we confront the notion that human evolution may not solely be a natural process but could also be influenced by external forces, potentially including extraterrestrial intervention. This perspective invites a reevaluation of our evolutionary path, suggesting that hybridization with alien species may have introduced new genetic material, thereby accelerating our development and adaptation in unforeseen ways.

One of the most significant implications of alien hybridization is the potential for expanded cognitive and physical capabilities. If extraterrestrial beings possess advanced genetic traits, their integration with human DNA could lead to enhanced intelligence, resilience, and adaptability. This raises questions about what it means to be human and whether the merging of species could result in a new form of existence, one that transcends our current limitations. Such advancements could redefine our understanding of intelligence and consciousness, prompting a deeper inquiry into the very nature of life itself.

The psychological effects of hybridization are another critical area of consideration. Individuals who identify as hybrids—or those who believe they have been subjected to alien genetic manipulation—often report a range of psychological experiences, from heightened sensitivities to a profound sense of otherness. These experiences can influence personal identity, societal interactions, and even the perception of spirituality. The implications extend beyond the individual, echoing through cultural narratives and societal beliefs about what it means to coexist with beings that may not adhere to our conventional understanding of humanity.

Ethical considerations surrounding alien genetics also emerge as a pressing concern. If hybridization is indeed a real phenomenon, questions about the morality of genetic manipulation arise. The potential for exploitation, consent, and the rights of hybrid beings must be critically examined. This discourse intersects with broader societal debates about genetic engineering in humans, as we grapple with the implications of playing god with our own DNA. The ethical dilemmas posed by hybridization compel us to confront our responsibilities not just to each other but to future generations who may inherit a world shaped by these extraordinary genetic legacies.

Finally, the cultural interpretations of hybrid beings in literature, film, and historical accounts serve as reflections of our collective anxieties and fascinations with alien intervention in human evolution. Sci-fi narratives often explore themes of hybridization, presenting both utopian possibilities and dystopian warnings. These stories shape our perceptions and understanding of hybrids, influencing public attitudes and beliefs about the reality of alien encounters. As we navigate these cultural landscapes, we must remain vigilant in discerning the line between fiction and potential truth, recognizing that the discourse surrounding human evolution is as much about our imaginations as it is about empirical evidence.

## ~ 10 ~

# SPIRITUAL PERSPECTIVES ON HYBRID EXISTENCE

### SPIRITUAL BELIEFS SURROUNDING HYBRIDS

Spiritual beliefs surrounding hybrids often delve into the intersection of the metaphysical and the biological, presenting a complex tapestry of ideas that challenge conventional understandings of existence. Many proponents within the UFO and conspiracy theory communities argue that hybrid beings—those born of both human and extraterrestrial lineage—embody a new form of consciousness. These beliefs often posit that hybrids possess enhanced spiritual abilities or heightened awareness, allowing them to bridge the gap between humanity and other dimensions of reality. This perspective invites a reexamination of what it means to be human, suggesting that the integration of alien genetics may unlock latent potential within the human psyche.

The notion of hybrids as spiritually evolved beings resonates deeply within various cultural narratives. Across different societies, myths of interbreeding between humans and otherworldly entities have long existed, often casting hybrids as mediators or messengers of divine knowledge. In many of these stories, hybrids are seen as holding a unique position that allows them to access higher realms of spiritual understanding and wisdom. This archetype is reflected in contemporary UFO lore, where hybrids are frequently described as possessing abilities such as telepathy, precognition, and other psychic phenomena, further enhancing their status as spiritually significant entities.

Furthermore, the spiritual implications of hybridization raise important ethical questions regarding the manipulation of human genetics. As hybrid beings are theorized to represent the convergence of different evolutionary paths, discussions around their rights and place within human society emerge. If hybrids are endowed with unique spiritual attributes, what responsibilities do we hold towards them? Such inquiries challenge not only our ethical frameworks but also our understanding of kinship, belonging, and the essence of spiritual evolution. This discourse is critical for enthusiasts and theorists alike, as it navigates the moral landscape of potential genetic engineering and the implications of creating life forms that transcend traditional definitions of humanity.

Psychological effects of hybridization also play a significant role in the spiritual beliefs surrounding these beings. Individuals who claim to have encountered hybrids or who feel a connection to them often report profound shifts in their consciousness, leading to an expanded sense of self and an enhanced spiritual journey. These experiences can catalyze transformative processes, prompting individuals to rethink their place in the universe and their relationship with the cosmos. Such psychological phenomena underscore the belief that hybrids may serve not only as physical entities but also as catalysts for spiritual awakening among those who engage with them.

Ultimately, the exploration of spiritual beliefs surrounding hybrids invites a broader dialogue about the nature of existence itself. As hybridization blurs the lines between human and alien, it challenges us to reconsider the spiritual dimensions of life and the potential for transcending our current understanding of reality. This ongoing exploration resonates deeply within the realms of UFO enthusiasts and conspiracy theorists, propelling them toward a greater inquiry into what it means to exist as a hybrid in a world filled with mysteries yet to be unraveled. The spiritual journey of hybrid beings may well reflect our collective quest for meaning in a universe that is far more complex and interconnected than we have previously imagined.

## THE CONCEPT OF SOUL IN HYBRIDS

The exploration of the concept of soul in hybrids invites a profound inquiry into the intersection of human consciousness and extraterrestrial influence. At the core of this discussion lies the question of whether hybrid beings, the product of human-alien genetic manipulation, possess a soul akin to that of their human progenitors. This inquiry is not merely philosophical; it has tangible implications for our understanding of identity, morality, and the essence of being. Proponents of hybrid theories argue that these entities, created through advanced genetic engineering, may embody a new form of consciousness, potentially different from traditional human experience.

Cultural interpretations of hybrid beings further complicate the understanding of their souls. In many mythologies and religious narratives, hybrids often serve as bridges between realms, encapsulating divine and earthly qualities. From ancient tales of demigods to modern depictions of alien-human hybrids, the representation of these beings often reflects humanity's fears and aspirations. The notion that hybrids could possess an evolved or enhanced soul challenges conventional perceptions of spirituality and consciousness, suggesting that they may experience existence in ways beyond human comprehension.

Psychological effects associated with hybridization also play a crucial role in shaping the soul's narrative. Individuals claiming to have undergone hybridization often report profound psychological transformations, including altered perceptions of self and reality. These experiences raise questions about the impact of alien genetics on human psychology and whether such alterations might extend to the spiritual realm, potentially leading to new forms of awareness and understanding. The implications for those involved in these experiences are significant, as they navigate a world that often dismisses their claims as mere delusions.

The ethical implications of alien genetics cannot be overlooked in this discussion. If hybrids possess a soul, what rights and considerations should be afforded to them? This query delves into the moral responsibilities of those engaging in genetic engineering, whether human or extraterrestrial. It challenges the boundaries of what it means to be sentient and deserving of dignity. As society grapples with the ramifications of genetic manipulation, the recognition of a hybrid soul could necessitate a reevaluation of legal and ethical frameworks governing such beings.

Lastly, the exploration of hybrids in sci-fi literature and film has profoundly influenced public perception of their souls. These narratives often reflect societal anxieties about technology, identity, and the unknown. They serve as cautionary tales and sources of fascination, shaping the collective consciousness regarding what it means to be a hybrid. As enthusiasts delve into these stories, they find a rich tapestry of interpretations surrounding the essence of hybrids, offering insights into humanity's ongoing struggle to define its place in a rapidly evolving universe.

## ~ 11 ~

# GOVERNMENT CONSPIRACY THEORIES REGARDING HYBRIDS

ALLEGATIONS OF GOVERNMENT COVER-UPS

The notion of government cover-ups surrounding alien encounters and hybridization has persisted for decades, captivating UFO enthusiasts and conspiracy theorists alike. Central to this discussion is the belief that governments have withheld critical information about extraterrestrial life and their alleged experiments with human genetics. Numerous whistleblowers, from former military personnel to scientists, have claimed that they witnessed or were involved in clandestine operations aimed at studying or even creating hybrid beings. These allegations raise significant questions about transparency, accountability, and the ethical implications of government involvement in genetic manipulation.

Evidence supporting claims of government cover-ups often hinges on declassified documents, witness testimonies, and unexplained phenomena. For example, the infamous Project Blue Book, which investigated UFO sightings from 1952 to 1969, has been cited as a façade that masked deeper, more troubling explorations into alien technology and hybridization. Skeptics argue that the government's lack of thorough public disclosure creates a fertile ground for speculation, allowing conspiracy theorists to flourish. The tension between official narratives and public curiosity about extraterrestrial life fuels ongoing debates about how much information is being kept from the public and what the true motivations behind such secrecy might be.

Additionally, the psychological effects of hybridization claims on individuals involved cannot be overlooked. Many who believe they have been subjected to alien experimentation report profound psychological trauma, which can manifest as anxiety, depression, or even dissociation. This phenomenon raises ethical concerns regarding the treatment of these individuals and the potential exploitation of their experiences for governmental agendas. The idea that human-alien hybridization could be a reality not only challenges existing paradigms of human evolution but also forces society to confront the implications of such genetic engineering on identity and humanity itself.

Cultural interpretations of hybrid beings in literature and film further amplify the intrigue surrounding these allegations. Sci-fi narratives often explore themes of alien-human interactions, raising ethical questions about the manipulation of life forms and the potential consequences of such actions. These stories serve as a reflection of societal fears and hopes regarding the unknown, often paralleling real-life allegations of cover-ups and secret experimentation. The way these hybrids are portrayed can influence public perception, shaping beliefs about the possible existence of such beings and the moral dilemmas associated with their creation.

As discussions about government conspiracies regarding hybridization continue to evolve, it is crucial to consider the broader implications of these theories. They intersect with historical accounts of alien encounters, medical research into hybrid physiology, and even spiritual perspectives on existence. The convergence of these narratives creates a complex tapestry of belief and inquiry into what it means to be human in a universe that may harbor other intelligent life forms. Ultimately, the allegations of government cover-ups invite a deeper examination of the ethical, cultural, and scientific dimensions of hybridization, urging society to engage thoughtfully with the mysteries that lie beyond our understanding.

## EXAMINATION OF DOCUMENTED CASES

The examination of documented cases related to human-alien hybridization reveals a complex tapestry of anecdotal evidence, scientific speculation, and cultural narratives. Notable accounts from individuals who claim direct contact with extraterrestrial beings often include descriptions of hybridization experiences. These narratives typically encompass themes of abduction, experimentation, and subsequent psychological impacts, suggesting a profound intersection between human experience and perceived alien intervention. Such cases, while often criticized for their lack of empirical support, offer a rich field for exploration in understanding the broader implications of hybridization narratives within society.

Research into documented cases of alleged hybridization has often focused on a select group of individuals. These accounts frequently detail encounters with beings that exhibit both human and extraterrestrial characteristics, sometimes described as possessing elongated limbs or unusual features. The testimonies of these individuals frequently highlight experiences of forced medical procedures or genetic manipulation, raising questions about consent and the ethical ramifications of such actions. This body of evidence serves as a foundation for discussions surrounding the moral obligations of both humanity and potential extraterrestrial civilizations in the realm of genetic experimentation.

The psychological effects reported by individuals claiming hybridization experiences are varied and profound. Many describe feelings of isolation, confusion, and a sense of being caught between two worlds. Some have reported a strong sense of purpose or mission related to their experiences, which can lead to complex emotional and psychological outcomes. Such effects warrant further investigation into the mental health implications of these encounters, particularly as they relate to the broader themes of identity and belonging in a rapidly changing world. The narratives surrounding hybridization often challenge conventional understandings of human psychology and the nature of reality itself.

Culturally, the implications of documented hybridization cases extend beyond individual experiences, touching upon societal fears and fascinations with the unknown. These accounts resonate within the larger framework of science fiction literature and films, which frequently explore themes of alien-human interaction and genetic manipulation. By examining how these narratives are represented in popular culture, one can glean insights into societal attitudes toward the potential for alien intervention in human evolution. Additionally, the portrayal of hybrid beings often reflects deeper anxieties regarding technology, identity, and the future of humanity.

Finally, the intersection of documented cases of hybridization with government conspiracy theories opens another layer of exploration. Many enthusiasts and theorists suggest that these hybridization programs are not only real but are being actively concealed by governmental entities. This belief in a hidden agenda fosters a climate of distrust and speculation, further complicating the conversation around the ethics of genetic engineering—both terrestrial and extraterrestrial. As we delve into these documented cases, it becomes increasingly crucial to navigate the fine line between skepticism and belief, recognizing that the implications of hybridization reach into the very fabric of our understanding of existence and our place in the cosmos.

## ~ 12 ~

# MEDICAL RESEARCH ON HYBRID PHYSIOLOGY

### HEALTH IMPLICATIONS OF HYBRIDIZATION

The health implications of hybridization, particularly in the context of human-alien interactions, warrant rigorous examination given the potential for profound biological and psychological impacts. As theories surrounding the existence of hybrid beings gain traction among UFO enthusiasts and conspiracy theorists, it becomes imperative to consider how the genetic amalgamation of human and extraterrestrial DNA may influence overall health. The introduction of foreign genetic material could lead to unforeseen physiological changes, including immune responses that are atypical or even detrimental. Such alterations could manifest as increased susceptibility to diseases, novel health challenges, or even the emergence of entirely new syndromes that current medical science is ill-equipped to address.

Psychological effects are equally significant when discussing the health implications of hybridization. The act of being part human and part alien may result in complex cognitive and emotional experiences. Individuals who identify as hybrids might struggle with their sense of identity and belonging, potentially leading to psychological distress, anxiety, or depression. The dual heritage could create a rift between their human experiences and their alien attributes, fostering feelings of isolation or alienation from both communities. Furthermore, the societal stigmas surrounding hybrid beings could exacerbate these psychological challenges, pushing them further into the margins of human experience.

Ethical considerations surrounding the manipulation of human and alien genetics raise fundamental questions about the morality of such practices. The potential for creating hybrids that possess unique abilities or resistances may lead to a societal divide, where hybrids are either revered or reviled. The implications for healthcare access, genetic discrimination, and the rights of hybrid individuals are profound. If hybrids are seen as superior or inferior based on their genetic makeup, this could instigate a new form of social hierarchy, posing ethical dilemmas that society must confront. The intersection of genetics and ethics in this context is not merely theoretical; it has real-world consequences for how hybrid individuals might be treated in healthcare systems and beyond.

Culturally, the interpretations of hybrid beings vary widely across different societies, influenced by folklore, mythology, and contemporary narratives in science fiction literature and film. These cultural lenses shape public perception of hybridization, often oscillating between fear and fascination. The portrayal of hybrids in popular media can either humanize them or demonize their existence, affecting how society views and interacts with those who may claim hybrid identities. This cultural dynamic plays a crucial role in influencing health outcomes for hybrids, as societal acceptance or rejection can directly correlate with their psychological well-being and social integration.

The historical accounts of alien encounters, often characterized by abduction narratives, often include themes of hybridization. These experiences raise questions about the medical implications for those who report such encounters. If hybridization is indeed occurring, it opens up avenues for investigating the physiological adaptations or anomalies that may arise from such interactions. Understanding the health implications of hybridization requires a multidisciplinary approach, incorporating insights from evolutionary biology, medical research, and even spiritual perspectives. As the dialogue surrounding hybrid beings continues to evolve, so too must our understanding of the complex health implications that arise from this intriguing intersection of humanity and the unknown.

ADVANCES IN MEDICAL RESEARCH

Advances in medical research have significantly expanded our understanding of genetic engineering and manipulation, particularly in the context of human-alien hybridization. Researchers are increasingly exploring the genetic material derived from unidentified biological entities, seeking to uncover the secrets of their physiology and potential applications for human health. This exploration raises questions about the intersection of advanced genetics and the boundaries of ethical medical practices. As scientists delve deeper into the genetic codes believed to be influenced by extraterrestrial life, they may unlock new pathways for treating diseases, enhancing human capabilities, and even addressing aging.

The psychological effects of hybridization are also becoming a focal point in medical research, with studies examining the mental and emotional impacts on individuals who believe they have undergone hybrid experiences. These investigations often reveal a complex interplay between belief systems, trauma, and identity, suggesting that the experience of hybridization may lead to profound psychological transformations. By understanding these effects, researchers can develop better mental health interventions tailored for individuals grappling with their experiences, whether real or perceived. The implications of these studies extend beyond individual well-being, prompting broader societal discussions about the nature of reality, consciousness, and human identity.

Ethical considerations surrounding alien genetics are paramount as advancements in medical research continue to challenge existing paradigms. The prospect of integrating alien DNA into human genomes raises profound questions about consent, identity, and the potential for unforeseen consequences. Researchers must navigate a labyrinth of moral dilemmas, balancing the potential benefits of hybridization against the risks of exploitation and dehumanization. The discussions surrounding these ethical implications are crucial, as they shape the policies and regulations governing genetic experimentation and the treatment of hybrid beings.

Cultural interpretations of hybrid beings also play a significant role in shaping public perception and acceptance of medical advancements related to human-alien interactions. From ancient folklore to contemporary science fiction, hybrids have been depicted as both saviors and threats, influencing how society views the potential of genetic manipulation. These narratives can either foster curiosity and acceptance or incite fear and resistance, affecting funding and support for research initiatives. Understanding the cultural context in which hybridization is discussed allows researchers to communicate their findings more effectively and engage with communities resistant to the implications of their work.

Lastly, the exploration of hybridization within the realms of science fiction literature and film provides a unique lens through which medical research can be interpreted. These narratives often reflect societal anxieties and aspirations regarding genetic engineering, serving as both cautionary tales and blueprints for potential futures. As medical research continues to push the boundaries of what is possible, the stories we tell about hybrids will evolve, influencing public discourse and shaping the future of human-alien relations. Ultimately, the interplay between advances in medical research and cultural narratives will determine how society navigates the complex landscape of genetic alchemy, particularly in the context of human-alien hybridization.

## ~ 13 ~

# CONCLUSION AND FUTURE DIRECTIONS

SUMMARY OF KEY FINDINGS

The exploration of human-alien hybridization reveals a complex interplay between genetic engineering and the broader implications of such interactions. Key findings indicate that the phenomenon of hybrid beings is not merely a fringe topic within UFO lore but a subject steeped in historical accounts, psychological effects, and ethical dilemmas. Various cases documented throughout history suggest that encounters with extraterrestrial entities often report experiences that align with themes of genetic manipulation. These narratives indicate a potential lineage of hybridization that raises questions about the origins and purposes of these beings, suggesting that such interactions may have been occurring for centuries.

Research into the psychological effects of hybridization points to profound impacts on individuals who believe they have been involved in such experiences. Reports frequently highlight feelings of isolation, confusion, and a sense of being different or chosen. This psychological landscape is further complicated by societal perceptions and stigmas surrounding UFO phenomena, which can lead to a lack of support for those who describe their encounters. The emotional and mental health consequences of these beliefs underscore the need for a nuanced understanding of how hybridization narratives affect individuals and communities, as well as the implications for broader societal acceptance.

Ethical implications surrounding alien genetics present a formidable challenge. The manipulation of human genetics, whether by extraterrestrial influence or advanced technology, raises significant moral questions. The potential for creating hybrid beings blurs the lines of what it means to be human and poses dilemmas regarding consent, agency, and the rights of such entities. As the boundaries of science and ethics are tested, a critical examination of these issues becomes necessary, particularly as they relate to governance and societal norms concerning hybrid beings.

Cultural interpretations of hybrid beings reveal a rich tapestry of beliefs and narratives that span across different societies and eras. From ancient mythologies to contemporary science fiction, these interpretations reflect humanity's fascination with the unknown and the desire to understand our place in the cosmos. Literature and film have played a pivotal role in shaping public perceptions of hybrids, often portraying them in ways that reflect contemporary anxieties about technology, identity, and the future of humanity. Such cultural artifacts not only entertain but also provoke thought and discussion about the implications of hybrid existence in our evolving understanding of life and consciousness.

In conclusion, the investigation into hybridization serves as a lens through which to explore fundamental questions about evolution, identity, and our relationship with the universe. The intersection of science, ethics, and cultural narratives provides a multifaceted view of what it means to be part of a potentially hybridized future. As we continue to uncover the mysteries of alien engineering and its implications, it is essential for UFO enthusiasts and conspiracy theorists alike to engage with these findings critically, fostering a dialogue that bridges the gap between speculative theories and empirical research.

### FUTURE RESEARCH AND EXPLORATION

Future research and exploration in the realm of genetic alchemy and alien engineering promises to unveil profound insights into the complex tapestry of human-alien hybridization. Scientific advancements in genetic engineering, such as CRISPR technology, pave the way for potential breakthroughs in understanding both the biological and psychological dimensions of hybrid beings. By investigating the genetic markers that may differentiate hybrids from their human progenitors, researchers can explore the implications of such modifications on health, cognitive abilities, and emotional responses. This inquiry not only enriches our comprehension of genetics but also raises critical questions about the boundaries of human identity and the ethical ramifications of manipulating life on such a fundamental level.

The psychological effects of hybridization remain an underexplored area ripe for investigation. Anecdotal accounts from individuals who claim to have experienced hybrid encounters reveal a spectrum of emotional and cognitive consequences, including altered perceptions of reality, identity crises, and heightened sensitivities. Future studies could focus on the psychological profiles of individuals who identify as hybrids or who have undergone alien encounters, providing valuable data on the mental health impacts of these experiences. By employing rigorous psychological methodologies, researchers can contribute to a more nuanced understanding of how the integration of alien genetics might influence human psychology over generations.

Ethical implications surrounding alien genetics must also be a focal point of future discourse. As genetic engineering becomes increasingly sophisticated, questions about consent, the rights of hybrid beings, and the moral responsibilities of those who engage in such research become paramount. Establishing ethical frameworks that govern the study and potential manipulation of hybrid genetics is crucial to prevent abuses and to protect the rights of all entities involved. By engaging ethicists, legal theorists, and scientists in a collaborative dialogue, we can ensure that the advancements in this field are pursued with caution and respect for both human and alien life.

Cultural interpretations of hybrid beings provide another rich avenue for exploration. From ancient mythologies to contemporary science fiction, hybrids have been portrayed in myriad ways that reflect societal fears, hopes, and ethical dilemmas. Future research could analyze how these narratives shape public perception of hybridization and influence policymaking regarding alien encounters and genetic experimentation. By examining literature, film, and folklore, scholars can uncover the evolving cultural landscape surrounding hybrids and contribute to a broader understanding of how these narratives affect real-world attitudes toward genetic engineering and alien life.

Finally, the intersection of hybridization and evolutionary biology offers fertile ground for future exploration. Understanding how hybridization might influence evolutionary trajectories could reshape our comprehension of human development and adaptability. Studies could investigate the potential for hybrids to thrive in diverse environments, examining their physiological traits and survival strategies in comparison to both human and alien counterparts. By integrating insights from evolutionary biology with data gathered from hybridization research, scientists can develop a comprehensive picture of how these beings might challenge or complement existing evolutionary theories, potentially leading to groundbreaking revelations about the future of our species and its place in the cosmos.